The Paper Maker

The Paper Maker

How to make handmade paper from fruits and vegetables

Ellaraine Lockie

COLLINS & BROWN

This book is dedicated to my papermaking students over the years, from whom I've learned as much as I've taught.

First published in Great Britain in 2001 by
Collins & Brown Limited
London House
Great Eastern Wharf
Parkgate Road
London SW11 4NQ

9 8 7 6 5 4 3 2 1

British Library Cataloguing-in-Publication Data: A catalogue record for this book is available from the British Library

ISBN: 1-85585-821-5

Editors: Jane Ellis, Michelle Pickering, Ulla Weinberg
Designers: Liz Brown, Alison Lee
Photographer: Sian Irvine
Indexer: Dorothy Frame

Color reproduction by Classic Scan Limited, Singapore
Printed by C&C Offset Limited, Hong Kong

This book was typeset using TheSans and Weiss

Note: Detailed safety guidelines for making paper are given on pages 8–9. However, the author, publisher, and copyright owner accept no responsibility for any damage or injury caused or sustained while using the products or techniques outlined in this book.

contents

introduction

Handmade paper is so much more than just paper. It is craft, art, therapy, meditation, gift, and tactile and visual pleasure. As well as being beautiful to look at and pleasurable to touch, handmade paper is also immensely versatile to use. It is perfect for letters, bookmarks, origami, lamp and candle shades, business cards, gift paper and tags, and much more.

Historically, handmade paper has been used as a conveyor of prayers, for offerings to gods, for ritual foldings, for books on magic, and to honor ancestors. In some cultures, it is considered to be as fine as silk, and in others, the hand-papermaking process is regarded as a divine activity. A handmade sheet of paper has been known to inspire a man's first love letter, to end long-term disputes, and to provide the medium for award-winning art.

The emphasis in this book is on how to make papers for stationery using inedible parts of fruits and vegetables. These papers are often highly textured, but always have one side that is satiny smooth for writing. However, the basic pulp can also be used to make paper for art and craft purposes.

This book is written for beginner papermakers, giving simple but comprehensive directions and avoiding the terminology used by professional hand papermakers. However, because it teaches a successful and dependable method of making strong paper from weak fruit and vegetable fibers, the book is also useful for expert papermakers.

These papers are easy and inexpensive to make, using everyday kitchen equipment and material that would otherwise be thrown in the garbage. Using inedible parts of fruits and vegetables not only creates exquisite papers with intriguing textures and variations, but also saves the earth's natural resources and is an example of recycling at its finest. For those of you who garden, there are even more possibilities, as you can make good papers using stalks, vines, and foliage from nearly any garden fruit or vegetable.

The techniques taught in this book can also be used to make strong paper from recycled paper products, as well as from partially processed sheets of fiber available from papermaking suppliers.

Opposite: *The vibrant colors of these cornhusk papers were achieved by airbrushing the dried papers with ink.*

safety

Although papermaking is one of the safest crafts you can do at home, there are potential hazards associated with some of the materials and processes. Read the safety guidelines outlined here, and take care that you always follow them.

Alum: Used for mordanting pulp in preparation for coloring (see page 21). Some people may be allergic to alum, but no special precautions are needed when using the recommended aluminum acetate or ammonium aluminum sulfate.

Clean up: Thoroughly clean the work area with a sponge or wet mop after you have used dyes, pigments, or potentially toxic materials. Do not use a vacuum, since small particles will go through the filters.

Colorants: Commercial dyes and some natural dyes (such as madder) can be toxic, so use appropriate gloves and respiratory protection. Be careful not to lift powders into the air. For airbrushing, only use coloring agents labeled for that use, and work in a well-ventilated area.

Cooking precautions: Cook fibers outside on an electric hotplate. Do not breathe the fumes when cooking.

Electrical safety: Keep your work area as dry as possible, and install ground-fault circuit interrupters on outlets near sinks and hotplates.

Equipment: Never use items that are utilized for food preparation.

Ferric ammonium citrate: Used for sunprinting (see page 120). This is not very toxic, but should not be ingested or inhaled. Provide ventilation or wear an appropriate face mask.

Gloves and eye protection: Wear rubber gloves (or plastic gloves if you are allergic to rubber) and vented splash goggles when using chemicals.

Hydrogen peroxide: Used for lightening paper (see page 20). The type for household use can burn the skin and eyes if it is in prolonged contact.

Material safety data sheets: Request a safety data sheet from the manufacturer for each product you use, and follow the advice about handling and storage.

Potassium ferricyanide: Used for sunprinting (see page 120). This can release hydrogen cyanide gas if exposed to strong ultraviolet light, strong acids, or high heat (higher than the heat from household ovens). It can be explosive when mixed with certain chemicals. Store away from other chemicals and out of sunlight, and wear appropriate gloves and respiratory protection.

Respiratory protection: If airborne contaminants cannot be avoided, provide ventilation to carry them outdoors. Use a mask that seals to your face, and make sure it is appropriate for the materials you are using.

Soaking precautions: Some organisms in standing water can cause disease. They are killed when fibers are cooked, but some people can suffer allergic reactions. Change water regularly, clean up spills promptly, wash hands after contact, and soak fibers in areas where there is no food or food preparation.

Solvents: All solvents are flammable, toxic, and irritating to the skin and eyes. Wear appropriate gloves and eye protection, and do not inhale. Store away from open flames, heat, or sparks.

Storage: Separate cooked fibers from food in the refrigerator or freezer, and label them clearly. Store all chemical products in accordance with manufacturer's safety data sheets.

TSP (trisodium phosphate): Used for sunprinting (see page 120). This is irritating to the skin and eyes. Wear appropriate gloves and eye protection.

Washing soda (sodium carbonate): Used for cooking fibers (see page 16). This is a mild alkali but is corrosive to the skin, eyes, and respiratory tract. Use ventilation or a mask if dust is raised, and wear appropriate gloves and eye protection. Always add washing soda to water before the water boils to avoid spatters. Never cook fibers with washing soda in aluminum, tin, or iron pots. The pots react and produce a gas. Instead, use stainless steel or enamel-coated pots.

Work area: Do not work with chemicals in areas used for food.

Protective gloves, goggles, and face mask for working with chemicals and colorants.

equipment & supplies

Most of the materials needed for papermaking can be found in hardware or kitchen supply stores. A list of papermaking supply companies can be found on pages 126–127. Most have catalogs and supply products by mail order.

Always use equipment made from stainless steel, glass, wood, or unchipped enamel for papermaking. Never use aluminum, tin, or iron because they can cause unattractive brown spots on the paper. Use different papermaking pots and utensils from the ones you use for food, and store separately. If you wish to embellish your paper, refer to the appropriate technique on pages 108–125 for the equipment needed.

You can use everyday kitchen equipment for papermaking.

preparing the fiber

- Inedible parts of fruits and vegetables (fresh, frozen, or dried) or recycled paper products (see box on page 19)
- Garden shears or kitchen scissors
- Strainer to rinse uncooked fibers
- Electric hotplate
- 21-quart (20-liter) standard canning pot (see box on page 12)
- Small cooking pot
- Measuring jug, cups, and spoons
- Appropriate protective gloves and safety glasses (see page 8)
- Washing soda or soda ash
- Stirring spoon
- Strainer bucket made from a 5-gallon (19-liter) plastic bucket, a large nylon paint strainer or other type of mesh, and three clothespins (follow the directions on page 12; see also box on page 12)

making the pulp

- Partially processed fiber or paper pulp to strengthen the fruit and vegetable fiber (abaca fiber—from the leafstalk of banana trees—can be purchased in sheets from papermaking suppliers, with ⅓lb (150g) of abaca sufficient for about 40 thin, stationery-size paper sheets

Stainless steel pots for preparing the fiber, and meat baster for embellishing the finished paper (see page 110).

when combined with fruit/vegetable fiber; see also page 18)
- One or two medium-size bowls
- Blender (do not use a food processor)
- 5-gallon (19-liter) plastic bucket (see box on page 12)
- Liquid starch (sizing to make the paper water-repellent and bleed-resistant; buy it or follow the directions on page 12)

forming the sheets

- Newspapers and plastic sheet to protect work surface from water (optional)
- Vat or tub (at least 7in (17.5cm) deep, with an area large enough for your mold and deckle to fit inside easily; a large, plastic cat litterbox or toy box is ideal)
- Mold and deckle (buy a set or follow the directions on page 13 to make your own)

Blender for making the paper pulp.

- Thick, absorbent towels
- Spray bottle
- Large laminated wooden board with a smooth surface for drying papers, such as tile board, or thick acrylic sheet, such as Plexiglass
- High-quality, absorbent sponge
- Toothbrush to clean mold screen
- Straight-edge razor or other thin blade
- Paper towels and books or weights to weigh down damp papers (old telephone books are excellent, and can also be used for pressing flowers and foliage)

Cooking pot and plastic buckets: You can use a smaller pot than 21 quarts (20 liters) if you wish, but you must adjust the amount of washing soda that you add to the fibers during the cooking process accordingly. You may also need to add small amounts of water to keep the fiber from thickening and burning; do so a little at a time to prevent boiling over. If you use a smaller cooking pot, you can use a smaller strainer bucket and blend a smaller quantity of pulp.

making liquid starch

To make enough for a 5-gallon (19-liter) bucket of well-sized pulp, stir ¼ cup of cornstarch into ¼ cup of cold water until dissolved. Bring 3¾ cups of water to a boil. Gradually add the cornstarch mixture while stirring. Simmer and continue stirring for two minutes until the mixture is lump-free. Use immediately, or reheat to the boiling point right before using.

making a strainer bucket

Use a strainer bucket to rinse cooked fiber outside with a garden hose. Drill three 1in (2.5cm) holes around a plastic bucket, evenly spaced and about two-thirds of the way up from the base. Hold the nylon paint strainer or other type of strainer mesh inside the bucket with clothespins.

Strainer bucket for rinsing cooked fiber.

making a mold and deckle

A mold is a frame with a screen over it used to form a sheet of paper. A deckle is a frame exactly the same size as the mold (minus the screen) that is placed on top of the mold to prevent the fiber from running off. It produces handmade paper's characteristic deckled edges.

You will need the following equipment:
• Wooden stretcher bars from an art-supply store or ¾in (2cm) plywood and waterproof glue
• Sandpaper (if you use plywood)
• Waterproofing sealant
• Brass strainer screen with 30 or 40 mesh from a plumbing or hardware store (do not use screen that is for screen doors)
• Kitchen shears
• Staple gun and staples
• Duct tape

Mold and deckle for forming paper sheets.

directions

1 Use waterproof glue to assemble wooden stretcher bars into two frames. The space inside the frames determines the size of your paper. Alternately, you can make stronger frames from plywood. The wooden edges should be about 1in (2.5cm) wide. Sand the edges lightly, waterproof them with a sealant, and let them dry.

2 With kitchen shears, cut a piece of brass strainer screen about 2in (5cm) larger than the inside of the frames. Wear gloves to protect yourself from the sharp edges. Use a staple gun to attach the screen to one of the frames. Start by putting one staple in the center of the north side, then have a second person pull tightly on the screen while you put a second staple in the center of the south side. Next, put one staple in the center of the east side, then have your helper pull tightly on the screen while you put the next staple in the center of the west side. Now put staples about every 2in (5cm) all around the screen, alternating sides and having your helper pull tightly on the screen each time. It is important to have the screen very tightly stretched.

3 Cover the wood on the screen side with duct tape. Over time the screen may start to sag. If this happens, simply replace the duct tape and screen instead of making a brand-new mold and deckle.

making
paper

preparing the fiber

Cellulose is the main ingredient in paper, so the non-cellulose parts of inedible fruit and vegetable pieces must be removed before you can use them to make strong, durable paper. This is accomplished by cooking them with washing soda.

The fiber of most inedible fruit and vegetable pieces breaks down easily, but preparation and cooking times vary. The variations for individual fruits and vegetables are given in the Paper Directory. If you want to make paper using only recycled paper products or sheets of processed fiber or pulp from papermaking suppliers, omit this process and go directly to pages 18–19.

directions

1 Cut the inedible parts of fruits and vegetables into 2in (5cm) pieces using garden shears or kitchen scissors. You can use fresh, frozen, or dried produce, but if you dry it, make sure it does not mildew. Rinse the pieces if there is excess debris, and soak them in water for 24 hours if you desire a shorter cooking time. You can omit the cutting and soaking if you cook them for longer than recommended.

2 Put the fruit and vegetable pieces in a 21-quart (20-liter) cooking pot (see box on page 12 if using a smaller pot), filling it half full if the produce is fresh or frozen, or a third full if it is dried. Add water to the three-quarters mark. In a small cooking pot, heat 1 cup of water. Before the water

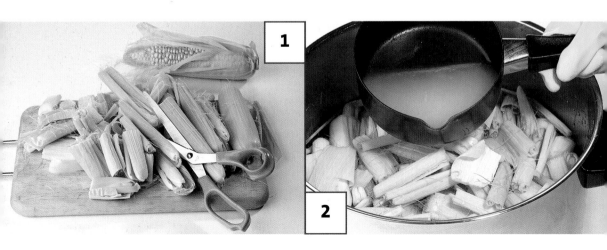

boils, add ½ cup of washing soda to it, stirring until the washing soda is dissolved. Stir the solution into the large pot.

3 Put the large cooking pot on an electric hotplate outside. Cover the pot and bring the fiber to a simmer. Simmer, stirring occasionally, until the fiber is mushy and slightly slippery, usually 2–5 hours after it begins to simmer. Dried fruit and vegetable pieces need to cook longer than frozen or fresh ones. To test if the fiber is ready, take some out with a spoon and, wearing appropriate protective gloves, give it a gentle tug. If it tears easily with and against the grain, it is ready. If not, continue cooking until it does, checking every half hour. Excessive overcooking (as in several hours too long) can damage fibers and produce a weaker paper, but precise timing is not crucial. Remove the pot from the heat and allow it to cool.

4 In a strainer bucket (see page 12), rinse the fiber for 2–3 minutes. Do this outside using a garden hose if possible. Remove the nylon strainer containing the pulp from the bucket and squeeze out the water. You can do this on the pavement by stepping on the pulp with your feet while wearing waterproof shoes. Put the squeezed fiber and nylon strainer back in the bucket. Rinse and squeeze until the water runs clear, usually three or four times. This rinsing is important so that all the non-cellulose material is removed. The fiber is now ready to be made into pulp. It can also be labeled and refrigerated in a plastic bag for up to a week, or frozen indefinitely.

> **Safety:** Follow the safety guidelines on page 9 for working with washing soda, always cook the fibers outside on a hotplate, and do not breathe the steam.

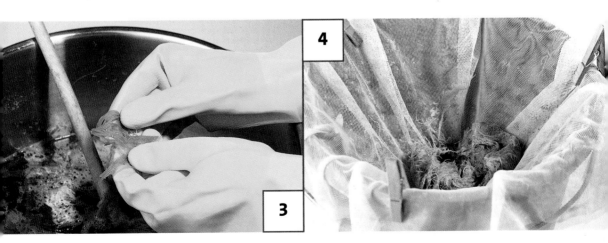

making the pulp

Fruit and vegetable fibers are generally shorter and weaker than those in other plants, and produce fragile papers if used alone. However, by adding a stronger fiber such as abaca, you can make beautiful, thin, and exceptionally strong paper.

Abaca—fiber from the leafstalk of a banana tree—is available in the form of sheets of partially processed fiber that simply needs hydrating before use. The amount of abaca needed for each of the fruit and vegetable fibers is given in the Paper Directory. There are other types of processed fiber or pulp available if you prefer, or if you are unable to obtain abaca. Cotton fiber, known as "cotton linters," is one option. It will produce thicker and weaker paper than abaca, but works better than abaca for embossing. You can also use recycled high-quality papers instead of abaca, though again, the sheets will not be as strong.

directions

1 Soak around ⅓lb (150g) of abaca in a bowl of water for at least half an hour. Tear it into 1in (2.5cm) pieces. This will be sufficient to make around 40 sheets of paper, when combined with fruit and vegetable fiber.

2 Put a handful of torn abaca pieces (¼ –⅓ cup) into the blender and fill two-thirds with water. Keep the area around the blender as dry as possible to avoid electrical shock. Run the blender on high speed for 30 seconds, briefly stopping every 10 seconds to rest the motor. Add

a handful of fruit or vegetable fiber (¼–⅓ cup) that has been cooked, rinsed, and squeezed. Run the blender for another 5–30 seconds; recommended times are given in the Paper Directory. Precise timing is not critical, but it is important to run the blender long enough to macerate the fibers evenly, but not so long that you shorten the fibers and weaken them. Do not run the blender for more than a total of 60 seconds for each blender-load.

3 Test whether the pulp has been blended sufficiently by putting a teaspoon of pulp into a jar of water. Cover the jar and shake it. If the fibers are reduced to an equal fineness, they are the right consistency. If there are clumps left, it needs to be blended longer. Pour the blended pulp into a large bucket and repeat steps 1 and 2 until all the cooked fruit or vegetable fiber has been blended.

4 Size the pulp to make it water-resistant by stirring in 2 cups of concentrated commercial liquid starch or 4 cups of homemade liquid starch (see page 12). Leave the pulp to stand for at least half an hour before forming the sheets. This amount of sizing is appropriate for stationery; papers intended for other purposes may need less or no sizing at all. You can add a few drops of essential oil to the pulp to scent the paper if you wish.

Making pulp without fruits or vegetables: Follow the same directions, but replace the plant fiber with paper products such as newspapers, telephone directories, and egg cartons. Run the blender for 10–15 seconds after adding the paper product. Alternately, use partially processed fiber sheets or high-quality paper such as envelopes, greeting cards, and photocopy paper, running the blender for the full 60 seconds continuously.

coloring the paper

Fruit and vegetable fibers usually have color of their own, which sometimes remains in the fibers after cooking, but it tends to be very subtle. You can increase the color range of your papers using commercial or natural dyes, or a combination of colorants.

This book illustrates only a few of the many beautiful colors that can be achieved. Experimentation is the key to success, particularly when using natural dyes, since these are more unpredictable than commercial ones. After lightening or coloring the pulp with any of the methods described, run it through the blender again (5–10 seconds per blender-load) to make sure you get paper with an even texture.

lightening the pulp

Some fruit and vegetable fibers produce very dark paper—banana peel and wild mushrooms, for example. You can lighten these fibers or strip the color completely from naturally light fibers by soaking the pulp in a hydrogen peroxide solution.

To do this, pour 1–2 cups of hydrogen peroxide into a large bucket of pulp. Stir well, then leave in a cool place out of direct sunlight until the pulp reaches the desired degree of lightness. This can take 1–5 days. Rinse the pulp thoroughly in a strainer bucket (see page 12) before coloring or forming paper sheets.

mordanting

Most natural substances and some commercial products used to color pulp require that the pulp first be soaked in a mordant. A mordant is a substance that allows the dye color to be fixed to the pulp. Instructions on whether mordanting is required are given for each of the dyeing methods described.

The only mordant used for the papers in this book is alum in the form of ammonium aluminum sulfate, used in home pickling and obtainable at drug stores, or alum in the form of aluminum acetate, a better choice because it works for more types of fibers and produces stronger colors. It is available from dye suppliers.

1 Dissolve ½ cup of alum into 1 pint (470ml) of hot water. Stir this into a 21-quart (20-liter) cooking pot containing one batch of pulp. Adjust the quantity of alum for smaller amounts of pulp. Leave for a few hours or overnight.

2 Rinse the pulp well in a strainer bucket (see page 12), then return the pulp to the cooking pot. Fill the pot with fresh water, then color the pulp with any of the dye methods that require a mordant.

built-in mordants

Some natural substances, such as saffron and turmeric, have a built-in mordant.

Simply stir the colorant into the pulp before forming the sheets. Start with 1tsp (5ml) and keep adding until you achieve the required shade.

Other substances with built-in mordants require a more lengthy method to achieve strong colors—coffee, tea, and henna, for example. For these, add the substance to boiling water, then turn off the heat and leave for a few hours or overnight. Use this dyed water to make the pulp in the blender. Let the pulp sit for a few hours or overnight before forming the sheets.

> **Safety:** Read the safety instructions on page 8 regarding the use of alum, hydrogen peroxide, and colorants. Lighten, mordant, and color papers in an area away from food or food preparation, wear appropriate protective gloves and face mask, and work outside or in a well-ventilated area.

Below, left to right: *Lightened banana peel; onion skins colored with hot-water fabric dye; carrot pulp colored with ink; artichoke leaves colored with cochineal; cornhusks airbrushed with food coloring; broccoli stalks colored with soil.*

pigments and dyes

To color a bucket of pulp with commercial pigment or dye, follow the manufacturer's instructions, then strain the pulp as usual. Pigments usually need a retention agent; both can be purchased from papermaking suppliers. Cold- and hot-water fabric dyes come with a mordant in each package.

paper napkins

Add 1–2 solid-color paper napkins to each blender-load of pulp during the last 5–10 seconds of blending. The resulting colors will be a shade or two lighter than the napkins. Mordanting is not necessary.

inks and paints

Inks, watercolors, and acrylics produce beautiful colors when added to mordanted pulp. Add the desired amount of color, then heat the pulp to just below boiling point for about 1 hour. Strain as usual.

airbrushing

Inks, watercolors, and acrylics that are labeled for airbrush use can be airbrushed onto the surface of dried paper. Do this outside or in a well-ventilated area. Fill a perfume atomizer three-quarters full with water, then add the color with an eye-dropper until you have the desired intensity. Shake the atomizer and spray the color directly onto dried sheets of paper.

Cover the sheets with paper towels and books for a couple of hours to prevent them from curling, then let the sheets air-dry. Mordanting is not required.

food and food coloring

Food colorings and foods such as spirulina are another source of dyes. First, mordant the pulp. Add small amounts of dye to 2 cups of boiling water. Pour 1 tbsp (15ml) into the pulp, stirring continuously and gradually adding more until you achieve the color you want. Heat to just below boiling point for about 1 hour, then remove from the heat. Leave for several hours or overnight, then strain as usual.

fruit/vegetable water

Many fruit and vegetable fibers lose most of their color when cooked in washing soda. To retain some of the natural color, do not add washing soda for the first hour of cooking. Strain the water and reserve it, then add fresh water with washing soda to the fiber and finish cooking. Rinse well, then mordant the pulp. Add the reserved water to the pulp and bring to just below boiling point. Turn off the heat, leave for several hours or overnight, then strain.

Opposite: Tea bags, paper napkins, food coloring, inks, watercolors, acrylics, and natural dyes in the form of concentrated powders and extracts are just a few of the many ingredients that can be used to color handmade paper.

plants and insects

Certain flowers, leaves, stems, twigs, barks, roots, wood shavings, and insects are sources of natural dyes. Put the dyestuff in an appropriate-size pot and fill it two-thirds full with water. Bring it to a boil and simmer for 1 hour. Strain off the liquid and reserve it, then follow the directions for dyeing with fruit/vegetable water.

powders and extracts

Natural dyes such as cochineal, logwood, madder, quebracho, chestnut, cutch, pomegranate, osage orange, fustic, and indigo come in concentrated powder and extract forms (see Resources, page 126). Some need to be made into a paste before use, while others have to be dissolved in hot water first; follow the manufacturer's directions. When you have prepared the powder or extract, follow the directions for dyeing with food and food coloring.

soil

Use clay soil with a rich, deep color. Put a layer of dirt in a large glass jar, then fill it with hot water. Leave it in bright sunshine for at least 1 hour, then pour off the colored water and reserve it. Repeat with fresh dirt and water until you have enough colored water. Strain the water to remove dirt and tiny pebbles. Mordant the pulp, then follow the directions for dyeing with fruit/vegetable water.

forming paper sheets

Fruit and vegetable fiber has a high shrinkage rate as it dries, causing it to crinkle when using traditional sheet-forming and drying methods. The revolutionary technique described here is foolproof and will help you make beautiful, flat papers.

directions

1 Place newspapers around the vat to soak up dripping water if you wish. Fill the vat about two-thirds full of water and add 8–10 cups of pulp. The thickness of a sheet is determined by the proportion of pulp to water; the more pulp you put in the vat, the thicker the paper will be. Use more pulp if you are making paper that does not use abaca or some other type of partially processed fiber. Each time you form a new sheet, add another 1–3 cups of pulp. Once you have figured out the thickness you want, just keep adding that amount of pulp before forming each sheet.

2 Wet the mold's screen by dipping it into the vat. Position the mold, screen-side up, in one hand and place the deckle over the top of the mold. With your other hand, stir the pulp in the vat with your fingers spread wide, in a lengthwise direction only, until the pulp is well dispersed (15–20 seconds this first time, but only a few seconds for each subsequent sheet; if you stop making sheets for any length of time, however, you must stir the vat water for 15–20 seconds again before making the next sheet). It is important to do this stirring only in a lengthwise direction so that water does not splash out of the vat.

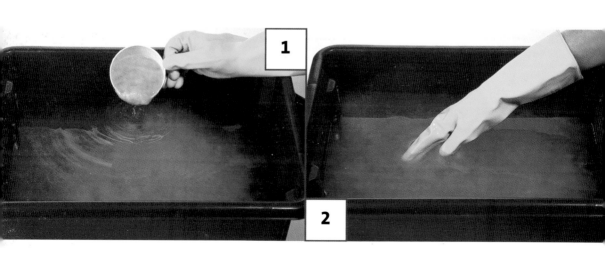

The pulp needs to be thoroughly stirred because it tends to settle on the bottom of the vat, and if the fibers are not well dispersed, the paper will be too thick in some places and too thin in others.

3 Holding the mold and deckle in both hands, dive them, submarine-style, to the bottom of the vat; the leading edge should be almost vertical when entering the water, before leveling the mold and deckle into a horizontal position when completely submerged. Keeping them horizontal, bring them slowly to the surface again. They will have a thin layer of wet fibers on them. Give them four quick little shakes. Each shake should be a quick tilt—only 1–2in (2.5–5cm) in each direction—from side to side lengthwise, and from side to side widthwise. Do not tilt any more than this because the fiber may slide on the screen. These shakes are important

because they make the wet fibers interlock in all directions to form strong sheets of paper. Let the excess water drain.

4 When most of the water has drained, rest the mold and deckle on the edge of the vat with one hand. Remove the deckle with your other hand. When removing the deckle, lift it straight up from the mold, helicopter-style, and keep it absolutely level while moving it across the mold. This is important so that water does not drip from the deckle and leave water spots that will be visible on the dried paper. If at any time you want to throw a wet sheet back into the vat water and start over, just place the mold, screen-side down, on the water surface, and the pulp on the screen will release into the vat. Stir the vat of pulp water again (in a lengthwise direction) to disperse the fibers before starting over, but now stir for only a few seconds.

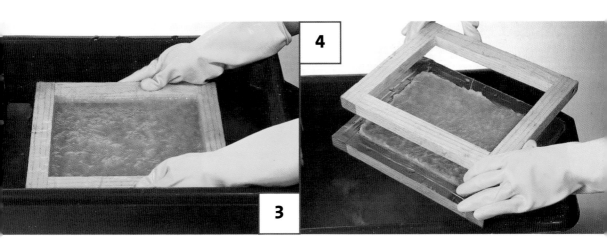

5 Take the mold off the edge of the vat and place it on a flat surface, fiber-side up. Fold a towel in half, gently place the double-thick towel over the fiber on the screen, and lightly press across its surface. Do not drag the towel over the fibers or press too hard; the idea is to take just enough water out of the fiber so that it is strong enough to come off the mold successfully. The more water that is removed from the fiber, the stronger the paper becomes. Do not towel-blot more than once, however. If you remove too much water at this stage, the sheet will dry too quickly when laid on the drying board, causing the sheet's edges to crinkle. This is the most important step when making paper with weak fibers.

6 With a spray bottle, spray a paper-size section on the drying board liberally with water. Flip the mold onto the wet board, fiber-side down. The board needs to be wet so that the paper dries without crinkling. Sponge water out of the sheet with an absorbent sponge by firmly pressing the sponge against the back of the screen, squeezing the water out of the sponge into a bowl as you work. Go over all areas of the sheet with the sponge 4–5 times. Remove the mold from the board by lifting it from one of its shorter ends, leaving the sheet of fiber on the board. Gently press the sheet with the sponge another couple of times, removing more water and pressing out any air bubbles.

> **Clean-up:** Do not pour unused pulp water down the drains because it may clog them. Strain the pulp out of the water using a strainer bucket (see page 12). Scrub the mold's screen with a toothbrush and running water after use, so that pulp does not clog the holes.

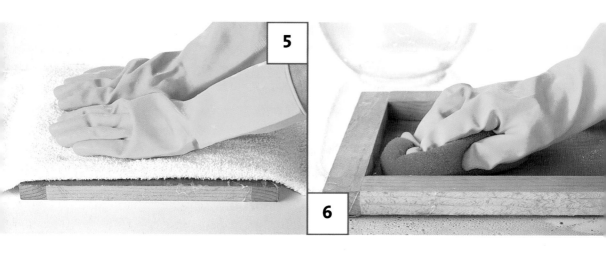

drying the sheets

Drying paper on a smooth, slick surface such as a tile board or sheet of Plexiglass produces one satiny smooth side (the side that touches the board) suitable for writing on, and one rougher, more textured side.

1 Leave the sheet on the drying board indoors, out of sunlight and drafts, until it is thoroughly dry. This will take 1–2 days, depending on the type of fiber, how thick the sheet is, how much humidity is in the air, and how warm the temperature is. The sheet needs to dry slowly so that it does not crinkle or warp. Drying slowly also allows the fibers to shrink naturally, which gives the paper greater folding strength.

2 When the paper is completely dry, carefully peel it from the board, starting with two corners at one end. If the edges stick, carefully use a one-sided razor or

other thin blade to loosen them. The paper should feel crisp after you remove it. If it feels limp, it is still too damp and should have been left to dry longer on the board. (If this happens, finish drying the sheet between paper towels with a heavy book or other weight on top.) Stack the dry papers together and place a heavy book or other weight on top of them for a few hours to make sure that they stay flat.

scenting the paper

If you want scented paper but have not added essential oil to the pulp (see page 19), you can add the scent after the papers have completely dried. Soak a cotton ball in several drops of essential oil and put it in a box big enough to hold the papers. Put the papers in the box and cover the box. Leave for several days, or store the papers there indefinitely. The papers will retain a subtle but distinct scent.

drying the sheets

scenting the paper

whimsical notepapers

1 On lightweight cardboard or plastic needlepoint mesh a little larger than the screen on your mold, draw or trace a design in the shape that you want your paper sheet to be. The backs of notepaper pads are perfect for making notepaper templates. The design can be any size you wish, as long as it fits within the screen area. Cut out the design with scissors.

2 Spray the cardboard or plastic mesh surface that will be face-up on your mold with silicone spray (this will prevent pulp from sticking to the template). Center the template on the mold, then place the deckle on top of it. Make the sheet of paper in the usual way.

3 After sponging the sheet on the drying board and removing the mold, carefully peel off the cardboard or plastic mesh template. With your fingers or a chopstick, carefully push away the fiber around the edges of the design, leaving the fiber shape on the drying board. Toss the pushed-away fiber back into the vat of pulp.

4 Sponge the design to remove air bubbles, and dry the sheet of paper in the usual way. Rinse excess pulp from the cardboard or plastic mesh template by dipping it into the vat. You can use it again immediately or dry it for a later time.

easy envelopes

1 Find a commercial envelope of an appropriate size for your paper sheets. Unglue the envelope and spread it open. With a pencil and ruler, draw a rectangle that snugly fits the edges of the unfolded envelope. (Do not worry about the curves.) Make a mold and deckle following the directions on page 13, with the screen the same size as the measured rectangle of the unfolded envelope.

2 Use this mold and deckle to make a sheet of envelope paper in the usual way, but add extra pulp to the vat for each envelope sheet, since it needs to be fairly strong. When the envelope sheet is dry, lay it on a flat surface, and position the unfolded commercial envelope on top of it.

3 Fold the two together, following the creases of the commercial envelope.

4 With your handmade envelope folded, clip away any excess paper from the four corners with a pair of scissors. Glue the envelope's seams (special envelope glue is available from papermaking suppliers, though you can use ordinary glue if you wish). Sealing wax or a sticker is enough to hold the envelope together if it is not going through the mail. If you want a more substantial envelope, glue the commercial envelope inside the handmade one.

problems & solutions

If you follow the directions in this book, you should find papermaking simple and satisfying. However, things can go wrong from time to time, especially if you are new to the process, so here are some solutions to common problems.

problem: Ink sinks into the finished paper, or your pen does not glide smoothly across the paper's surface.
solution: Spray the completely dried sheets with spray starch and iron them dry between sheets of parchment paper.
prevention: Add more liquid starch to the bucket of pulp before forming sheets.

problem: A wet sheet has air bubbles in it after it has been flipped onto the drying board.
solution: Air bubbles can usually be worked out by pressing the sponge down on each bubble in a series of presses, moving the bubble along until it disappears at the end of the sheet.
prevention: Sponge more firmly and longer before you remove the mold. If this does not work, replace your sponge with a new, highly absorbent one.

problem: A wet sheet of paper has folds or tears in it after it has been flipped onto the drying board.
solution: Throw the sheet back into the vat, stir well, and form another sheet.

prevention: Press more firmly when you blot with the towel to remove water from the sheet. Or you may need a new screen on your mold if it has started to sag. Or the sheet may be too thin, for which you need to add more pulp to the vat.

problem: The fiber in the sheet is not evenly distributed. There are spots that are too thick or too thin.
solution: Throw the sheet back into the vat, stir well, and form another sheet.
prevention: Stir the pulp in the vat for a longer time in order to disperse the fibers more thoroughly. When forming the sheet in the vat, make sure you are taking the mold and deckle right to the bottom of the vat and leveling it before you bring it up again.

problem: There are lumps of fiber in the wet sheet.
solution: Throw the wet sheet back in the pulp bucket and macerate the pulp in the blender until the lumps are gone.
prevention: Macerate blender-loads of pulp for longer.

problem: A wet sheet is sticking to the mold or has holes in the same place for several sheets in a row.

solution: For sticking, towel-blot the sheet several times, peel it off the mold, and place it on the wet drying board. Cover holes with pulp using a meat baster.

prevention: Scrub the mold screen with a toothbrush and running water because it may be clogged with pulp.

problem: Sheets are not turning out as well as they have in the past.

solution: Your mold screen may have started to sag and should be replaced with a tightly stretched one. Or you may need to replace your sponge, since sponges lose their capacity to hold water over time.

problem: The edges of the paper start to curl in an unsightly manner as it dries on the drying board.

solution: If you notice the curling when it begins, cover the sheet with paper towels and a heavy book. Change the towels as they become damp. Repeat until the paper is dry to the touch, then air-dry for another few hours before removing the sheets from the drying board.

prevention: Spray more water onto the drying board before you flip the sheet onto it. Do not allow direct sunlight to hit the drying paper. Press less water from the sheet when you towel-blot it.

problem: A sheet is damp on the underside, even though it feels dry on the top, or it starts to curl after removing it from the drying board.

solution: Cover the sheet with paper towels and a heavy book for a few hours. Change the paper towels when they become damp.

prevention: Leave the sheets on the drying board for a few hours longer after the sheets feel dry.

problem: The ends of a wet sheet have flipped over on top of the sheet.

solution: You can sometimes use a thin blade to lift up the outer edge and move it back. If not, throw the sheet back into the vat and form another.

prevention: Be careful not to drag the towel over the wet sheet as you are blotting it. Sponge more water from the upside-down mold on the drying board before removing the mold.

problem: The pulp starts to smell bad.

solution: Strain the water from it, then add fresh water. Fruit and vegetable fiber pulp usually stays fresh-smelling at room temperature for about three days. It lasts longer in the refrigerator and indefinitely in the freezer.

prevention: Refrigerate the pulp if you are not going to use it within three days, or strain, label, and freeze it.

paper
directory

cornhusks

The husks and silk from corncobs make some of the finest handmade paper. Cornhusk paper is very strong, but looks and feels luxurious. It is especially well suited for stationery.

preparing the pulp

To prepare the cornhusks, follow the directions on pages 16–17, cooking the fiber for 4–5 hours. To make the pulp, follow the directions on pages 18–19, but put both abaca and cornhusks into the blender at the same time and use no more than ½ cup of fiber (both abaca and cornhusks together) for each blender load. You can use more cornhusks than abaca if you want a more textured paper, going as high as a 75:25 ratio. Blend for the full 60 seconds. If there is balled-up cornhusk fiber on the bottom of the blender, put it in the next blender load to disperse it.

coloring and forming the sheets

If you choose to color the pulp, use one of the methods on pages 20–23, or try some experiments of your own. To form the sheets, follow the directions on pages 24–26. To dry the sheets, follow the directions on page 27.

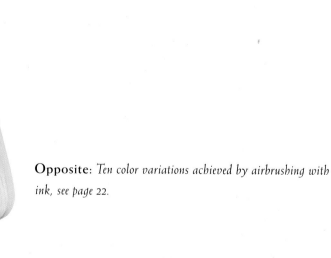

Opposite: *Ten color variations achieved by airbrushing with ink, see page 22.*

cornhusks

Colored with food coloring
SEE PAGE 23

Colored with turmeric
SEE PAGE 21

Natural-colored

Airbrushed with watercolor
SEE PAGE 22

cornhusks

Colored with cold-water fabric dye, with cotton lint from tumble dryer partially embedded with meat baster
SEE PAGES 22 & 110

Colored with cold-water fabric dye, with straw partially embedded with meat baster
SEE PAGES 22 & 110

Colored with chestnut bark extract, with sunflower petals added in blender
SEE PAGES 23 & 110

Natural-colored base sheet; top border sheet colored with soil and embossed with plastic stencil, then attached with spray adhesive
SEE PAGES 23 & 117

pineapple tops

Pineapple tops make luscious-looking paper with a slight variegated quality. It also has one of the smoothest and glossiest writing surfaces of all handmade papers.

preparing the pulp

To prepare the pineapple tops, follow the directions on pages 16–17, cooking the fiber for 4–5 hours. To make the pulp, follow the directions on pages 18–19, blending for 15–20 seconds after adding the pineapple fiber. You can use more pineapple fiber than abaca if you want a more variegated paper, going as high as a 80:20 ratio.

coloring and forming the sheets

If you choose to color the pulp, use one of the methods on pages 20–23, or try some experiments of your own. To form the sheets, follow the directions on pages 24–26. To dry the sheets, follow the directions on page 27.

Opposite, top: *Natural-colored.* **Center right:** *Colored with acrylic paint, with jute trim partially embedded with meat baster, see pages 22 & 110.* **Bottom right:** *Colored with ink, see page 22.* **Bottom left:** *Colored with concentrated watercolor and embossed with plastic stencil, see pages 22 & 117.* **Center left:** *Colored with pomegranate extract and embossed with crocheted doily, see pages 23 & 116.*

pineapple tops

Colored with
concentrated watercolor
SEE PAGE 22

Colored with ink
SEE PAGE 22

Colored with
pomegranate extract
SEE PAGE 23

Natural-colored

pineapple tops

Colored with cold-water
fabric dye, with seaweed
added in blender
SEE PAGES 22 & 110

Colored with concentrated
watercolor, with floating
abaca ribbon
SEE PAGES 22 & 118

Colored with pomegranate extract,
with dried flowers partially
embedded with small screen
SEE PAGES 23 & 111

Colored with cold-water
fabric dye and embossed
with crocheted doily
SEE PAGES 22 & 116

onion & garlic skins

Yellow onion skins produce wonderful tan, gold, and yellow papers, while red onion skins produce paper with a hint of purple. Garlic skin paper is a stunning, snowy white color.

preparing the pulp

To prepare the skins, follow the directions on pages 16–17, though you do not need to cut them into pieces. Fill the pot with skins, then pour hot water over them to condense them. As the volume in the pot decreases, fill it up with more skins, then add more hot water. Continue until the pot is two-thirds full. Cook garlic skins for 1–2 hours and onion skins for 3–4 hours. To make the pulp, follow the directions on pages 18–19. Blend onion skins for 15–20 seconds, going as high as an 80:20 ratio of skins to abaca. For garlic skin pulp, blend the abaca for 60 seconds, then add the skins and blend for just a few seconds longer; a 50:50 ratio of skins and abaca works well.

coloring and forming the sheets

If you choose to color the pulp, use one of the methods on pages 20–23, or try some experiments of your own. Onion skin pulp has a strong natural color, so commercial dyes are usually more successful than natural ones, though lightening the pulp before coloring can alleviate this problem. To form the sheets, follow the directions on pages 24–26. To dry the sheets, follow the directions on page 27.

Opposite, top left: *Garlic skins colored with hot-water fabric dye, see page 22.* **Top right:** *Yellow onion skins, lightened and then colored with concentrated watercolor, with yarn bits added in blender, see pages 20, 22, & 110.* **Bottom right:** *Red onion skins.* **Bottom left:** *Garlic skins with pressed flowers partially embedded with meat baster, see page 110.*

onion skins

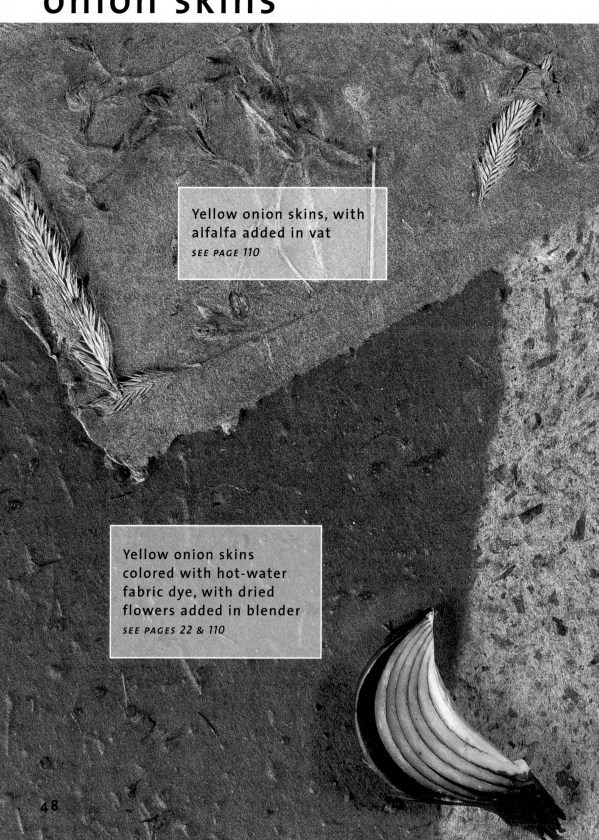

Yellow onion skins, with alfalfa added in vat
SEE PAGE 110

Yellow onion skins colored with hot-water fabric dye, with dried flowers added in blender
SEE PAGES 22 & 110

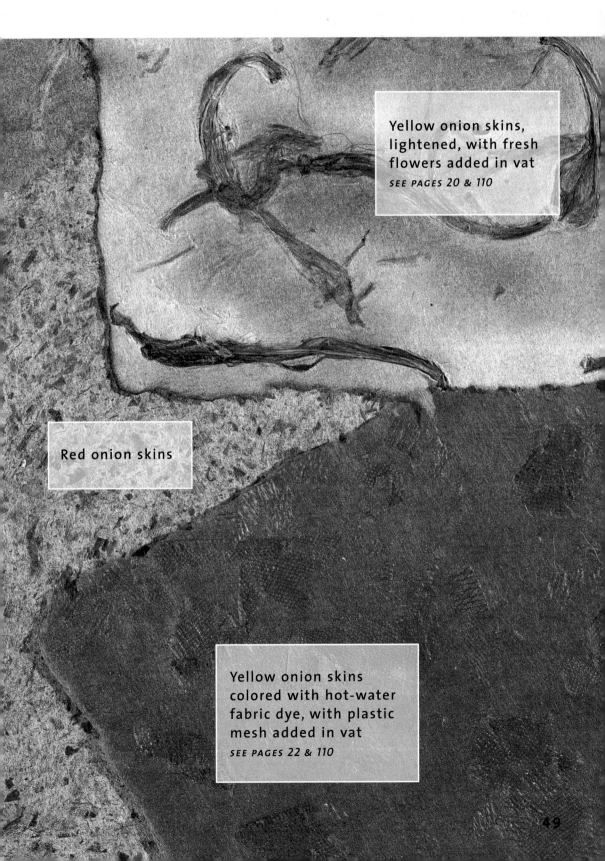

Yellow onion skins, lightened, with fresh flowers added in vat
SEE PAGES 20 & 110

Red onion skins

Yellow onion skins colored with hot-water fabric dye, with plastic mesh added in vat
SEE PAGES 22 & 110

garlic skins

Colored with food coloring
SEE PAGE 23

Colored with hot-
water fabric dye
SEE PAGE 22

Colored with cold-water fabric
dye, with weeds added in vat
SEE PAGES 22 & 110

Colored with cold-water fabric dye
SEE PAGE 22

Colored with food coloring
SEE PAGE 23

Colored with blue denim linters, with dried flowers added in blender and vat, and partially embedded with meat baster
SEE PAGES 18 & 110

Colored with food coloring, with leaves partially embedded with meat baster
SEE PAGES 23 & 110

broccoli stalks

Paper from broccoli stalks has a slightly rough texture that is very pleasing to the touch. The wet fiber is easy to handle, so this is a good paper for beginners to make.

preparing the pulp

To prepare the broccoli stalks, follow the directions on pages 16–17, but soak the stalks in water for several days before cooking. Cook the fibers for 4–5 hours, stirring more frequently than usual because broccoli fiber tends to burn more easily than other fibers. Do not worry if it browns on the bottom of the pot; this will simply give the paper an earthier tint. To make the pulp, follow the directions on pages 18–19, blending for about 30 seconds after adding the broccoli fiber. A 50:50 blend of abaca and broccoli fiber works best.

coloring and forming the sheets

If you choose to color the pulp, use one of the methods on pages 20–23, or try some experiments of your own. To form the sheets, follow the directions on pages 24–26. To dry the sheets, follow the directions on page 27.

Opposite, top left: *Natural-colored.* Top right: *Colored with spirulina, with clover leaves partially embedded with meat baster, see pages 23 & 110.* Bottom right: *Lightened, see page 20.* Bottom left: *Colored with spirulina, with cookie-cutter cutouts, see pages 23 & 124.*

broccoli stalks

Colored with cold-
water fabric dye
SEE PAGE 22

Colored with hot-water
fabric dye and spirulina
SEE PAGES 22 & 23

Colored with red soil
SEE PAGE 23

Colored with instant coffee
SEE PAGE 21

broccoli stalks

Colored with instant coffee,
with dried foliage partially
embedded with small screen
SEE PAGES 21 & 111

Colored with spirulina,
with cookie-cutter cutouts
SEE PAGES 23 & 124

Colored with cold-water
fabric dye, with dried flowers
added in blender and vat
SEE PAGES 22 & 110

Colored with spirulina,
with clover leaves partially
embedded with meat baster
SEE PAGES 23 & 110

melon rinds

Paper made from melon rinds has a leathery look and a sandpaper feel. All varieties of melon are perfect for art projects, and some can be used for writing paper as well.

preparing the pulp

To prepare the melon rinds, follow the directions on pages 16–17, cooking the fiber for 5 hours. Rinds from any type of melon can be used, or you can combine several varieties. Cantaloupe rinds have the roughest texture, followed by watermelon. The surface of honeydew rind paper is suitable for writing. To make the pulp, follow the directions on pages 18–19, blending for about 30 seconds after adding the melon rind fiber. You can use slightly more melon fiber than abaca if you wish, with around a 60:40 ratio.

coloring and forming the sheets

If you choose to color the pulp, use one of the methods on pages 20–23, or try some experiments of your own. To form the sheets, follow the directions on pages 24–26. To dry the sheets, follow the directions on page 27. Melon rind paper takes longer than usual to dry, and will curl readily if you remove it from the board before completely dry.

Opposite, top left: *Honeydew melon rinds colored with hot-water fabric dye, see page 22.* **Top right:** *Honeydew melon rinds colored with cold-water fabric dye, see page 22.* **Bottom right:** *Honeydew melon rinds colored with hot-water fabric dye, see page 22.* **Bottom left:** *Watermelon rinds with raffia added in blender, see page 110.*

melon rinds

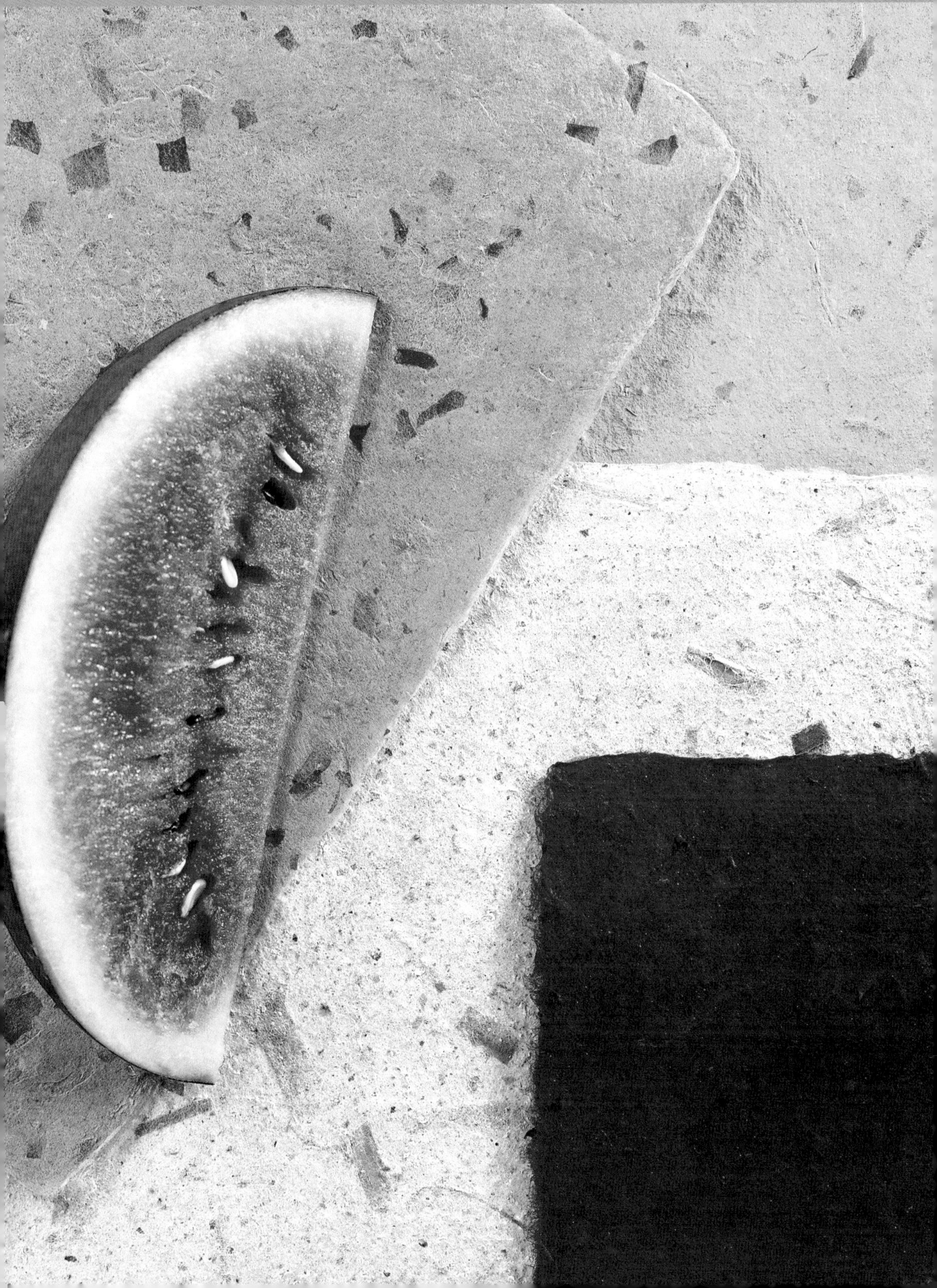

melon rinds

Honeydew melon rinds colored
with concentrated watercolor
SEE PAGE 22

Honeydew melon
rinds colored with
hot-water fabric dye
SEE PAGE 22

Watermelon rinds colored
with hot-water fabric dye
and wattle extract
SEE PAGES 22 & 23

Watermelon rinds with
used green tea leaves
added in blender
SEE PAGE 110

Honeydew melon rinds colored
with cold-water fabric dye
SEE PAGE 22

melon rinds

Watermelon rinds with
raffia added in blender
SEE PAGE 110

Honeydew melon rinds
colored with cold-water fabric
dye, with dried flowers added
in blender and vat
SEE PAGES 22 & 110

Watermelon
rinds with
potpourri
added in
blender
SEE PAGE 110

Honeydew melon
rinds colored with
hot-water fabric dye
SEE PAGE 22

Honeydew melon
rinds colored with
concentrated
watercolor, with
feathers partially
embedded with
small screen
SEE PAGES 22 & 111

artichoke leaves

Paper from artichoke leaves is strong, beautifully textured, and a good choice for embellishments that require hardy paper, such as sunprinting and marbling.

preparing the pulp

To prepare the artichokes, follow the directions on pages 16–17. Use all inedible parts of the artichoke head; separate the leaves and cut the other parts into 2in (5cm) pieces. Artichoke fiber will not break down to the extent required simply by cooking with washing soda, so you need to soak it in water for 5–6 weeks before cooking. Put the fiber in a bucket, cover with water, and put a lid on top. Change the water every few days (see soaking precautions on page 9). After soaking, cook the fiber for 5 hours. To make the pulp, follow the directions on pages 18–19, blending for about 30 seconds after adding the artichoke fiber. You can use more artichoke fiber than abaca if you want a more textured paper, going as high as a 70:30 ratio. If you use a high proportion of artichoke fiber, blend the abaca for a shorter time and blend longer after the artichoke fiber has been added. Altogether, blend for no more than 60 seconds.

coloring and forming the sheets

If you choose to color the pulp, use one of the methods on pages 20–23, or try some experiments of your own. To form the sheets, follow the directions on pages 24–26. To dry the sheets, follow the directions on page 27.

Opposite, top left: *With rose tea leaves added in blender, see page 110.* **Top right:** *Colored with hot-water fabric dye, with pressed flowers partially embedded with meat baster, see pages 22 & 110.* **Bottom right:** *Natural-colored.* **Bottom left:** *Colored with hot-water fabric dye, see page 22.*

artichoke leaves

Colored with
concentrated watercolor
SEE PAGE 22

Colored with
cochineal extract
SEE PAGE 23

Colored with hot-
water fabric dye
SEE PAGE 22

Natural-colored

artichoke leaves

Colored with hot-water fabric dye, with pressed flowers partially embedded with meat baster
SEE PAGES 22 & 110

With embroidery threads added in blender
SEE PAGE 110

Painted with flowers
SEE PAGE 115

**Colored with cold-water
fabric dye, with rose tea
leaves added in blender**
SEE PAGES 22 & 110

asparagus ends

Asparagus ends make strong paper with a fine, fibrous texture. It is excellent for sunprinting or marbling, since the dry sheets hold up well when rinsed in water.

preparing the pulp

To prepare the asparagus ends, follow the directions on pages 16–17, cooking the fiber for 4–5 hours. To make the pulp, follow the directions on pages 18–19, blending for about 20 seconds after adding the asparagus fiber. You can use more asparagus than abaca if you want a more textured paper, going as high as a 75:25 ratio.

coloring and forming the sheets

If you choose to color the pulp, use one of the methods on pages 20–23, or try some experiments of your own. To form the sheets, follow the directions on pages 24–26. To dry the sheets, follow the directions on page 27.

Opposite, top left: With dried parsley leaves added in blender, see page 110. Top right: Colored with hot- and cold-water fabric dyes, see page 22. Bottom right: Natural-colored. Bottom left: Colored with red quebracho extract, see page 23.

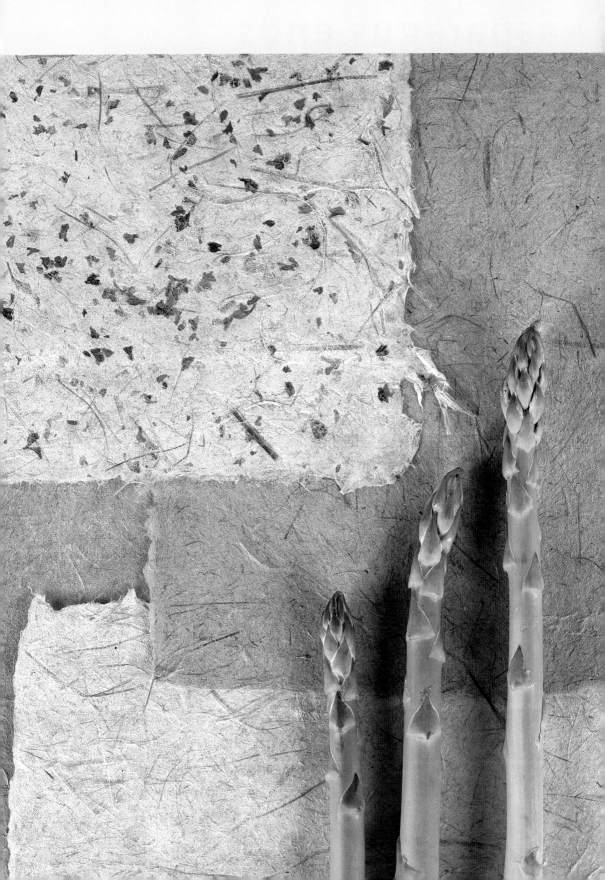

asparagus ends

Colored with hot-
water fabric dye
SEE PAGE 22

Natural-colored

Colored with cold-
water fabric dye
SEE PAGE 22

Colored with hot- and
cold-water fabric dyes
SEE PAGE 22

asparagus ends

Colored with red quebracho
extract, with pine needles partially
embedded with meat baster
SEE PAGES 23 & 110

With dried parsley
leaves added in blender
SEE PAGE 110

Colored with hot-
and cold-water fabric
dyes, with glitter
added in blender
SEE PAGES 22 & 110

Colored with cold-
water fabric dye and
red quebracho extract,
with dried rose
partially embedded
with small screen
SEE PAGES 22, 23, & 111

leek tops

Fiber from leek tops has medium-to-weak strength when wet, but produces pleasing paper with a smooth surface when dry. It works well for writing, embossing, and stenciling.

preparing the pulp

To prepare the leek tops, follow the directions on pages 16–17, cooking the fibers for 2 hours. Use only the green parts of the leeks, not the roots or stumps. Leek tops keep some of their color after cooking if you use them while they are fresh. When dried, they tend to lose their color during the cooking. To make the pulp, follow the directions on pages 18–19, blending for 10–15 seconds after adding the leek fiber. You can use slightly more leek fiber than abaca, with perhaps a 60:40 ratio.

coloring and forming the sheets

If you choose to color the pulp, use one of the methods on pages 20–23, or try some experiments of your own. The natural greenness in leek fiber can be combined with additional natural or commercial dyes to produce many complex colors. To form the sheets, follow the directions on pages 24–26. To dry the sheets, follow the directions on page 27.

Opposite, top: *Colored with logwood extract, see page 23.* **Center right:** *Colored with ink, see page 22.* **Bottom right:** *Colored with hot- and cold-water fabric dyes plus vegetable water reserved from cooking leek tops, see pages 22 & 23.* **Bottom left:** *Natural-colored.*

leek tops

Base sheet colored with acrylic paint; top laminated sheet natural-colored with floating flower
SEE PAGES 22, 112, & 119

Colored with myrobalon extract, with sawdust added in blender
SEE PAGES 23 & 110

Colored with hot- and cold-water fabric dyes, with feather partially embedded with meat baster
SEE PAGES 22 & 110

Colored with hot-water fabric
dye, with cotton lint from
tumble dryer partially
embedded with meat baster
SEE PAGES 22 & 110

Colored with acrylic paint,
with floating flowers
SEE PAGES 22 & 119

fibers from juicing

The leftover fiber, stems, and leaves from vegetables used for juicing make superb paper, as do the tops from vegetables such as carrots and beets.

preparing the pulp

To prepare the leftover fiber or vegetable tops, follow the directions on pages 16–17, but you do not need to cut anything into pieces. In addition to eliminating the chopping time, using this type of fiber also greatly reduces the cooking time, to just 1–2 hours. To make the pulp, follow the directions on pages 18–19, blending for 10–15 seconds after adding the fiber. A 50:50 blend of abaca and fiber works best.

coloring and forming the sheets

If you choose to color the pulp, use one of the methods on pages 20–23, or try some experiments of your own. To form the sheets, follow the directions on pages 24–26. To dry the sheets, follow the directions on page 27.

Opposite, top left: *Beet fiber from juicing colored with cold-water fabric dye, with fresh carnation blossoms added in blender, see pages 22 & 110.*
Top right: *Carrot tops colored with concentrated watercolor, with carrot tops added in blender, see pages 22 & 110.* **Bottom right:** *Natural-colored beet fiber from juicing.* **Bottom left:** *Carrot fiber from juicing colored with cold-water fabric dye, see page 22.*

fibers from juicing

Natural-colored carrot tops
base sheet; top border sheet
colored with concentrated
watercolor, then embossed
with plastic stencil and
attached to sheet beneath
with spray adhesive
SEE PAGES 22 & 117

Carrot tops with
pomegranate seeds
added in vat
SEE PAGE 110

Beet fiber from
juicing colored
with concentrated
watercolor and
embossed with
needlepoint mesh
SEE PAGES 22 & 117

Carrot fiber from juicing with
dried flowers partially
embedded with small screen
SEE PAGE 111

pumpkin shells

Pumpkin fiber is fairly weak when wet, but makes strong paper with interesting flecks when dry. Other squash shells can also be used, or different shells can be combined.

preparing the pulp

To prepare the pumpkin shells, follow the directions on pages 16–17. Use one large pumpkin shell or two medium-sized ones; jack-o'-lanterns left over from Halloween are ideal. Be sure to remove the seeds as you cut up the pumpkin, and do not use any of the stem, because both the seeds and the stem make hard, whitish clumps in the paper (there is no reason you cannot use these materials as embellishments, however). Cook for 2–3 hours. To make the pulp, follow the directions on pages 18–19, blending for 10–15 seconds after adding the pumpkin fiber. Keep the ratio about 50:50; too much pumpkin fiber will make weak pulp that is difficult to form into sheets.

coloring and forming the sheets

If you choose to color the pulp, use one of the methods on pages 20–23, or try some experiments of your own. Pumpkin shells keep their color fairly well after cooking and produce stunning, complex colors when combined with natural dyes. To form the sheets, follow the directions on pages 24–26. To dry the sheets, follow the directions on page 27.

Opposite, top left: *Colored with hot-water fabric dye, see page 22.* **Top right:** *Colored with two hot-water fabric dyes, see page 22.* **Bottom right:** *Natural-colored.* **Bottom left:** *Lightened, with leaf sandwiched between laminated sheets, see pages 20 & 112.*

pumpkin shells

Colored with two
hot-water fabric dyes
SEE PAGE 22

Colored with hot-
water fabric dye
SEE PAGE 22

Colored with cold-
water fabric dye
SEE PAGE 22

Colored with madder extract
SEE PAGE 23

Natural-colored

pumpkin shells

Colored with two hot-water
fabric dyes, with dried
flowers added in blender
SEE PAGES 22 & 110

With crushed autumn
leaves added in blender
SEE PAGE 110

Lightened, with leaf
sandwiched between
laminated sheets
SEE PAGES 20 & 112

Colored with vegetable water,
with yarn partially embedded
with meat baster
SEE PAGES 23 & 110

banana peel

Banana peel makes handsome, strong, dark paper flecked with intriguing, tiny dark dots. Do not worry if the banana peel has turned black; the paper will not be as dark.

preparing the pulp

To prepare the banana peel, follow the directions on pages 16–17, cutting off and discarding the hard knobs from the tops and bottoms of the peel. Cook the banana peel for 1–2 hours. To make the pulp, follow the directions on pages 18–19, blending for 15 seconds after adding the banana fiber. You can use more banana fiber than abaca, going as high as a 60:40 ratio.

coloring and forming the sheets

If you choose to color the pulp, use one of the methods on pages 20–23, or try some experiments of your own. Banana peel produces dark paper, so it is a good candidate for lightening (see page 20). You can get some interesting and unpredictable colors when dyeing banana peel pulp because of its natural pigmentation. To form the sheets, follow the directions on pages 24–26. To dry the sheets, follow the directions on page 27.

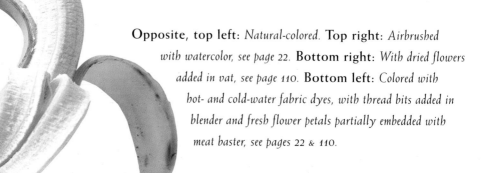

Opposite, top left: *Natural-colored.* Top right: *Airbrushed with watercolor, see page 22.* Bottom right: *With dried flowers added in vat, see page 110.* Bottom left: *Colored with hot- and cold-water fabric dyes, with thread bits added in blender and fresh flower petals partially embedded with meat baster, see pages 22 & 110.*

banana peel

Colored with cold-water fabric dye, with feathers added in blender and partially embedded with meat baster
SEE PAGES 22 & 110

Lightened, with poppy petals added in blender
SEE PAGES 20 & 110

Colored with hot- and cold-water fabric dyes, with thread bits added in blender and fresh flower petals partially embedded with meat baster
SEE PAGES 22 & 110

With dried flowers added in vat and partially embedded with small screen
SEE PAGES 110 & 111

citrus peel

Citrus peel fiber is weak when wet, but the dried paper is strong and has a splendid grainy surface. Any type of citrus peel will work, or you can combine peel from different fruits.

preparing the pulp

To prepare the citrus peel, follow the directions on pages 16–17. Remove seeds and stem knots before cutting the peel into pieces, and avoid using the inside membrane of fruits whenever possible. Cook the fiber for 2–3 hours. To make the pulp, follow the directions on pages 18–19, blending for 15–20 seconds after adding the citrus fiber. Keep the ratio about 50:50, because too much citrus fiber will make weak pulp that is difficult to form into sheets and to dry evenly.

coloring and forming the sheets

If you choose to color the pulp, use one of the methods on pages 20–23, or try some experiments of your own. To form the sheets, follow the directions on pages 24–26. To dry the sheets, follow the directions on page 27. Citrus peel has a non-fibrous texture that works well for embellishment techniques such as stenciling, embossing, and cookie-cutter designs. It is also perfect for scenting, because citrus essential oils are widely available (see page 27).

Opposite, top left: *Lime peel colored with vegetable water from carrot tops, with orange peel partially embedded with meat baster, see pages 23 & 110.* **Top right:** *Lemon peel colored with hot- and cold-water fabric dyes, see page 22.* **Bottom right:** *Grapefruit peel colored with cold-water fabric dye and embossed with plastic stencil, see pages 22 & 117.* **Bottom left:** *Orange peel colored with hot-water fabric dye, see page 22.*

citrus peel

Natural-colored
lemon peel

Lime peel colored with
hot-water fabric dye
SEE PAGE 22

Grapefruit peel
colored with
hot-water
fabric dye
SEE PAGE 22

Natural-colored
orange peel

Lime peel colored
with acrylic paint
SEE PAGE 22

Lime peel colored with
hot-water fabric dye
SEE PAGE 22

citrus peel

Grapefruit peel with
bougainvillea petals
partially embedded
with meat baster
SEE PAGE 110

Orange peel with plum
blossoms and leaves
added in blender
SEE PAGE 110

Grapefruit peel colored
with cold-water fabric
dye and embossed with
plastic stencil
SEE PAGES 22 & 117

Lime peel with
bougainvilla
blossoms and ferns
added in blender
SEE PAGE 110

Lemon peel with
fresh flower
blossoms added in
blender and pressed
flowers partially
embedded with
small screen
SEE PAGES 110 & 111

wild mushrooms

Wild mushroom paper is sensuous and silky to the touch, and the side that dries against the board has an unusually rich luster. Follow the safety precautions assiduously.

Safety: Some wild mushrooms are toxic or even fatal if eaten, and some wild mushroom cooking fumes are toxic if breathed. No wild mushroom species is harmful to handle, however, although you should not make this type of paper if you are allergic to mushrooms. Always use expert advice in identifying wild mushrooms, and take extra precautions by following these recommendations:

• Never store wild mushrooms or wild mushroom fiber in the refrigerator or freezer

• Take extra care not to inhale the steam when cooking wild mushroom fiber

preparing the pulp

To prepare the mushrooms, follow the directions on pages 16–17, cooking them for 4–5 hours. Some mushrooms will remain spongy and will not conform to the test for being done, but as long as the pieces are well cooked and well rinsed, they will make fine paper. To make the pulp, follow the directions on pages 18–19, blending for 20–30 seconds after adding the mushroom fiber. You can use more mushroom fiber than abaca, going as high as a 90:10 ratio. If you use a high proportion of mushroom fiber, blend the abaca for a shorter time, and blend longer after the mushroom fiber has been added.

coloring and forming the sheets

If you choose to color the pulp, use one of the methods on pages 20–23, or try some experiments of your own. Many mushroom species retain their natural color, and many types produce paper that is too dark to write on. These can be lightened by using more abaca or hydrogen peroxide (see page 20). To form the sheets, follow the directions on pages 24–26. To dry the sheets, follow the directions on page 27, making sure that you spray plenty of water on the board to prevent curling.

Opposite, top: *Golden mushrooms colored with cold-water fabric dye, see page 22.* **Bottom right:** *Cream-colored wild mushrooms.* **Bottom left:** *Golden mushrooms with dried flowers partially embedded with small screen, see page 111.* **Center left:** *Gray mushrooms with raffia added in blender and raffia bow partially embedded with small screen, see pages 110 & 111.*

wild mushrooms

Golden mushrooms with fresh
daisy petals added in vat
SEE PAGE 110

Golden mushrooms colored with cold-water fabric dye
SEE PAGE 22

Golden mushrooms colored with turmeric
SEE PAGE 21

Golden mushrooms with floating brazilwood sawdust
SEE PAGE 118

Gray mushrooms with raffia added in blender and raffia bow partially embedded with small screen
SEE PAGES 110 & 111

mixed fibers

Paper made from mixed fibers is always a visual and textural surprise. You can use a combination of the fibers in this book (see box), or you can experiment with other types.

Artichoke leaves:
Do not use artichoke leaves when making mixed-fiber paper, because the leaves must be soaked for six weeks before cooking. However, artichoke leaves that have already been made into pulp can be used.

preparing the pulp

To prepare mixed fruit and vegetable fiber, follow the directions on pages 16–17, cooking the fruit and vegetable pieces for 4–5 hours. A more convenient way to make mixed fiber paper is to save the last bit of pulp that you strain from the vat each time you make paper. Squeeze the excess water from it and store it in a clearly labeled bag in the freezer, adding to it as you go along. This pulp-accumulation method is perfect for a children's or teenagers' papermaking party. You can serve foods for dinner that have provided the papermaking fiber, then use the thawed pulp to make handmade paper. The paper can be colored with the paper napkins used for dinner. To make the pulp, follow the directions on pages 18–19, blending for 15–20 seconds after adding the fiber. A 50:50 blend of abaca and mixed fibers will produce an easy-to-handle pulp.

coloring and forming the sheets

If you choose to color the pulp, use one of the methods on pages 20–23, or try some experiments of your own. To form the sheets, follow the directions on pages 24–26. To dry the sheets, follow the directions on page 27.

Opposite, top left: *With paper confetti added in vat, see page 110.*
Top right: *Colored with paper napkins, see page 22.* **Bottom right:**
With crochet thread partially embedded with meat baster, see page 110.
Bottom left: *Colored with paper napkins, see page 22.*

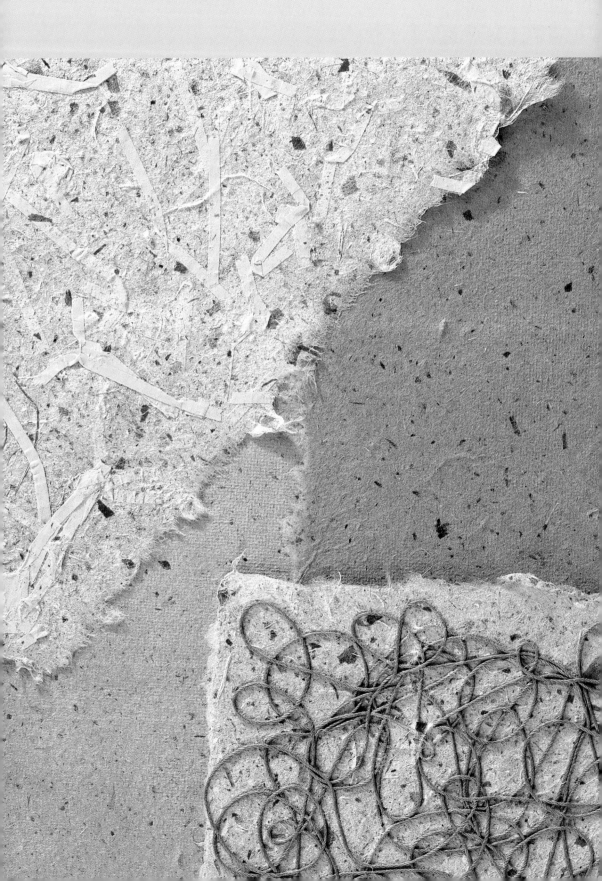

mixed fibers

With tissue paper
added in vat
SEE PAGE 110

Colored with paper napkins
SEE PAGE 22

Colored with paper
napkins and embellished
with floating string
SEE PAGES 22 & 118

Colored with
paper napkins
and embellished
with floating
embroidery thread
SEE PAGES 22 & 118

embellishing techniques

embedding

There are several ways in which you can embed materials in handmade papers to increase textural qualities, either adding them to the pulp or embedding them into the formed paper. Embedded papers can take up to twice as long to dry.

adding items to the vat

If you want paper with a rugged texture, add small bits of material to the vat of pulp before forming the sheets. You can use materials such as leaves, flowers, weeds, pine needles, feathers, moss, straw, animal hair, coffee, tea, spices, herbs, ribbons, yarns, threads, glitter, confetti, and so on.

adding items to the blender

If you want your paper to have some additional texture but require a relatively smooth surface for writing, add small bits of material in the final few seconds of

blending the pulp. Some additives are fragile and require only one short push on the blender button; others require 2–3 seconds of blending. For stationery purposes, soak the bits you want to add to the blender overnight so that they hydrate and are less likely to float on top of the pulp when it is in the vat.

partial embedding with meat baster

Use a meat baster to embed lightweight or fragile objects, such as pressed flowers. After sponging the sheet on the drying board, lay the item you wish to embed on

adding items to the vat

adding items to the blender

top of the sheet. Fill a meat baster with pulp water from the vat. Hold the baster about ½in (1cm) above the surface of the item and dribble the pulp all around the item's edge, being careful not to touch the item or the paper with the baster. Lightly sponge off excess water. Sometimes you will need to repeat the procedure or drip pulp water over the entire item, depending on the weight and size of the object.

The object will be a permanent part of the paper, delicately displayed when the paper is dry. If you do not use sufficient pulp water, or cover enough of the perimeter of the item so that it adheres completely after the paper is dry, touch it up with a little bit of glue.

partial embedding with small screen

Use a mini-screen to embed heavier items, such as twigs and stemmed flowers. Make a simple mini-mold by cutting a 2x3in (5x7.5cm) piece of brass plumbing screen. Tape a ½in (1cm) width of duct tape around the edges of the screen. After sponging a sheet on the drying board, place the item you wish to embed on top of the sheet. Dip the mini-mold into the vat just 2–3in (5–7.5cm), leveling it so that the mini-mold is covered evenly with a thin layer of pulp. Flip the mini-mold, pulp-side down, onto the sheet, covering a small part of whatever you wish to embed. Press out excess water from the screen's surface using a sponge. Carefully peel off the mini-mold, leaving the layer of pulp, and sponge the sheet of paper once more.

Part of the item will now be sandwiched between the main sheet of paper and the section you have applied with the mini-mold. This is a much more attractive method than simply gluing an item onto the paper surface.

partial embedding with meat baster

partial embedding with small screen

laminating

When two pieces of wet paper are pressed together, they fuse to become one piece. This is called laminating. It is used to make papers with different fronts and backs, and to sandwich objects between the sheets, creating a see-through effect.

Almost any flat, lightweight material can be sandwiched between two sheets of wet paper. Feathers, leaves, and pressed foliage work well, as do stenciled shapes and cookie-cutter designs (see pages 28 and 124). This technique is especially effective for lampshades because the sandwiched material shows through the light.

directions

1 Use a fiber that is light in color, or lighten it with hydrogen peroxide (see page 20). Garlic skin pulp works especially well, but many others are suitable if they are lightened in color. Make thin sheets so

there is a translucent quality when the paper is dry. Sponge the first sheet of paper on the drying board, then place the material that is to be sandwiched in the required position.

2 Make the top sheet of paper very thin by not adding any pulp to the vat for that sheet. Position the mold with the top sheet on it so that it exactly covers the sheet on the board. Sponge in the usual manner. A good way to line up the sheets is to mark the drying board so that you know where to place the mold each time. Finish the sheets in the usual way.

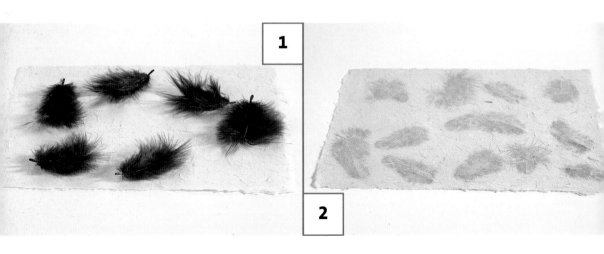

marbling

Marbling enhances the texture of handmade paper with dramatic and colorful designs. Use oil-based paint diluted with a little mineral spirits (white spirit), or an Easter egg coloring kit if available.

directions

1 Fill a tray that is slightly larger than the paper sheets with 3in (7.5cm) of cold water. Squeeze droplets of color across the surface of the water, and slowly drag a chopstick or skewer across it to swirl the colors into the desired pattern.

2 Lay a sheet of dry paper on top of the water for a few seconds until it is saturated. Do not immerse the sheet. Lift the sheet and transfer it to a drying board that has been sprayed with water. The sheet should be pattern-side up. If you wish, you can sprinkle some glitter onto the sheets at this point to produce a shimmering effect.

3 Before forming the next sheet, remove leftover colors from the water in the tray by placing a newspaper on the surface. The colors will adhere to the newspaper.

4 When the papers are partially dry, place paper towels and a book on top of each sheet to make sure that they do not curl. Replace the paper towels with dry ones

Safety: Wear appropriate protective gloves (see page 8), and do marbling outside or in a well-ventilated area.

every 30 minutes until the marbled papers are nearly dry. Remove the paper towels and the books, and allow the papers to dry completely before removing them from board. Clean your hands with vegetable oil, and clean the drying board with acetone or rubbing alcohol before using it for regular papermaking.

Marbling produces some of the most stunning handmade papers.

watermarking

A watermark is a translucent area in a sheet of paper that is barely visible except when held to the light. You can use duct tape, adhesive-backed paper, or lightweight cardboard to produce simple watermark designs.

using duct tape

Tear ½in (1cm) wide strips of duct tape, then press it firmly onto the dry screen in the required design. Alternately, use adhesive-backed paper to create the design. If the design is small, form and dry the sheet in the usual way. If the design is large, you may need to dry the sheet on the mold. Do this by blotting the sheet once with a towel after removing it from the vat. Then place the mold on top of some newspapers and let the sheet dry. If the edges start to curl, peel it from the mold, place it between paper towels, and cover with a book. When completely dry, peel the sheet off the mold. The duct-tape design can be reused several times.

using cardboard

Cut out the design from cardboard, and place it on the dry screen in the desired position. Hold the cardboard down with a thumb while you are immersing the mold in the vat. Release your thumb as you bring the mold up to the surface. The force of the water will hold the cardboard in place. Continue in the same way as when using duct tape, lifting the design off the sheet when both are dry. The design can be reused several times.

using duct tape

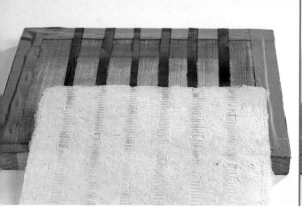

using cardboard

flower painting

Flower painting is like the finger painting that children like to do, except you use fresh flowers to create the vibrant, impressionistic designs. Unsized paper works better than paper with sizing for this technique.

painting on dry paper

Use flowers that leave color on the skin when pressed firmly between two fingers, such as begonias or marigolds, on completely dry paper. The center of some flowers, such as sunflowers, can be used to stain circular patterns. Test the leaves if you want green colors. Pistils from lilies also work well. When papers are still on the drying board, put a flower, stem, or pistil on your fingertip and "paint" a design with it, using as much pressure as you need to produce color. Leave the paper on the drying board for a few more hours, then remove it in the usual way.

painting on wet paper

Use flowers or petals that bleed color, such as lilies or lobelias, on top of wet paper while it is still on the drying board. Test flowers for bleeding by pressing one in an old telephone book for a few hours; if it leaves color on the pages, it is suitable for this technique. Use flowers within an hour of picking them, or freeze them in plastic bags ready for use later. When you have arranged the flowers in the required design, cover the sheet with plastic wrap and place a heavy book on top. Let the sheet dry, uncovered, before removing it from the drying board.

painting on dry paper

painting on wet paper

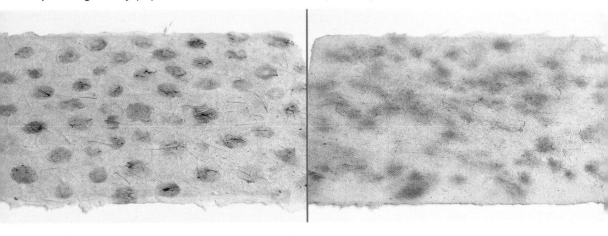

embossing & stenciling

You can make recessed or raised designs in your papers using a variety of materials. Plastic stencils create translucent impressions, while lace, crochet, plastic needlepoint mesh, and thin objects produce a more three-dimensional quality.

Embossed paper can take up to twice as long to dry as plain paper. Do not remove the objects until the paper is completely dry, because the embossed areas will tear easily if the paper is damp. If the finished sheet has too many holes—this can be a problem when using lace items and plastic stencils—use spray adhesive to attach it to a plain sheet of handmade paper. This can be very beautiful. If you want your finished embossed sheet to have no holes at all, use a meat baster to fill in the thinly covered areas with pulp from the vat after you have towel-blotted the sheet but before placing it on the drying board.

lace and crocheted items

Wet the lace or crocheted item, then wring out the excess water. Place the item in the desired position on the mold and cover with the deckle. Form the paper sheet in the usual way, using the deckle or your thumbs to hold the item in place. Leave the item in place until the paper is dry, then carefully peel it off before lifting the paper from the drying board.

plastic stencils

Use plastic stencils in the same way as lace or crocheted items, but gently scrape off

lace and crocheted items

plastic stencils

excess pulp from the top of the stencil with a chopstick after you have sponged the sheet on the drying board. This will allow you to lift off the stencil easily when the paper is dry. It also eliminates a "fringed" look around the stencil design.

needlepoint mesh

To create a smooth design with an embossed surrounding area, cut a shape in a piece of plastic needlepoint mesh that is slightly larger than the screen on your mold. Place the mesh on top of the mold and cover with the deckle. Make the paper sheet in the usual way, leaving the mesh on the sheet until the paper is dry. If you want a smaller sheet of embossed paper than the size of your mold, you can use a smaller piece of mesh and hold it in place with your thumbs rather than the deckle. Scrape away excess pulp from around the mesh with a chopstick, then leave to dry. To

create an embossed design with a smooth surrounding area, cut out a piece of mesh in the required shape, then hold it in place with both thumbs as you dip the mold and deckle into the vat. Leave the mesh on the sheet until the paper is dry.

thin objects

Choose thin items such as buttons, coins, and pressed leaves and foliage. Items that are no more than ¼in (6mm) thick work best. Spray the side of the object you wish to imprint on the paper with silicone spray to keep it from sticking, then put the item to one side. Make a sheet of paper in the usual way, but stop at the point where you would normally towel-blot it. Place the object in the desired position on the sheet. Towel-blot the sheet and proceed as usual, leaving the object on top until the paper is completely dry. Remove the object after removing the paper from the drying board.

needlepoint mesh

thin objects

floating materials

Natural fibers, pipe cleaners, and dried and pressed flowers and foliage can easily be bonded to handmade paper by "floating" them on the wet sheet of paper while it is still on the mold and before you towel-blot it.

floating fibers

1 Natural fibers in the form of yarn, thread, string, raffia, tissue paper, confetti, or printed words (from newspapers or non-glossy paper) are ideal for this technique. Synthetic materials do not bond as well, so use the partial-embedding technique with a meat baster or small screen (see pages 110–111) for those. Start by wetting your chosen material.

2 Position the wet fiber onto the wet sheet of paper when it is still on the mold and before you towel-blot it. Once in position, firmly blot and then sponge the

paper in the usual way. If the fibers are thin, it is not usually necessary to glue them in place after the paper has dried, as the fiber becomes part of the paper. If they are too thick, however, you may have to glue them after the paper has dried.

floating pipe cleaners

If you wish to float pipe cleaners, bend them into the desired shape the night before and dip them into the bucket of pulp so that they are evenly coated. You may need to dip them several times to cover them completely. Let them dry overnight. The next day, wet the

floating fibers

undersides of the pipe cleaners with pulp and press the wet sides onto the wet sheet of paper before towel-blotting it. This is especially useful for monogramming.

floating flowers and foliage

Press flowers and foliage in a telephone book within half an hour of picking them. Transfer them to another section in the book after a couple of days, to keep them from molding and permanently sticking to the pages while drying. You may like to carry a telephone book in your car for this purpose. Fix pressed flowers and foliage that are very thin and fragile to the surface of the paper by placing them on the wet sheet while it is still on the mold and before you towel-blot it. For thicker flowers and foliage, use the partial embedding technique with a meat baster described on page 110.

Form pipe cleaners into the required shape and soak in pulp overnight before floating them.

Floating yarn (left) and floating flowers (right).

floating pipe cleaners

floating flowers and foliage

sunprinting

Sunprinting, or blueprinting, is one of the oldest and most simple photographic processes. White images on a deep-blue background are created by placing objects on specially treated paper and exposing it to the sun.

Sunprinting should be done on a dry, sunny day with little or no breeze, and you should do the preparation in a room with no direct sunlight. The basic sunprinting process produces blue-toned prints, but you can alter the color to a yellow tone and then to a brown tone if you desire.

The paper must be rinsed after exposure to make the image permanent. Paper made from fruit and vegetable fibers holds up better in the rinsing process if you make the paper thicker than that used for writing. It is also best to use one of the stronger fibers, such as artichoke leaves, asparagus ends, or cornhusks.

You will need the following equipment:

- Appropriate respiratory protection (see page 9)
- Orange potassium ferricyanide and green ferric ammonium citrate (available from photographic and chemical suppliers, these should be stored in a lightproof plastic bag; see Resources, page 126)
- Scales or measuring spoons
- Two small paint rollers, 3–5in (7.5–12.5cm), depending on size of paper
- Plastic paint-roller tray of an appropriate size to accommodate rollers
- Hand-held hair dryer

Safety: Refer to the safety guidelines on pages 8–9 regarding the use of sunprinting chemicals, and always sunprint in a well-ventilated location away from food. When the chemically treated sheets are in strong sunlight, a tiny amount of hydrogen cyanide gas is released. If done outdoors, this is not significant, and it stops the instant the paper is removed from sunlight. Wear appropriate protective gloves when rinsing the sheets.

1

- 1in (2.5cm) deep sheet of foamboard slightly larger than the paper (available from art-supply stores; glue several sheets together to achieve the required depth if necessary)
- Items to form designs, such as lace, pressed foliage, and negative photograph images on acetate transparencies (can be produced at print shops)
- Straight pins (for attaching general items to foamboard)
- Piece of glass slightly larger than the paper (for holding acetate transparencies in place on foamboard)
- T-pins (large-headed pins that look like a letter T in profile for holding the sheet of glass in place)
- Two vats (one for rinsing blueprints, one for yellow- and brown-toned prints)
- TSP (for yellow- and brown-toned prints; available at hardware stores)
- Black teabags (for brown-toned prints)

blue-toned prints

1 First, make the sunprinting solution. Pour 8fl oz (250ml) of warm water into a container, then add 1tbsp (15ml) orange potassium ferricyanide and stir until it is dissolved. Gradually add 2tbsp (30ml) green ferric ammonium citrate, stirring after each addition. Stir the solution well.

2 Using a paint roller, apply some of the solution to each sheet of paper as evenly as possible. Use a second paint roller to even the solution if it puddles. Wipe the second paint roller with newspaper after each use, and keep it as dry as possible. Cover the solution whenever you are not using it, and stir it well before applying it to each sheet.

3 Blow the surface of the paper with a hair dryer until it is dry to the touch. Immediately place the sheet on a piece of foamboard and place the item you wish to

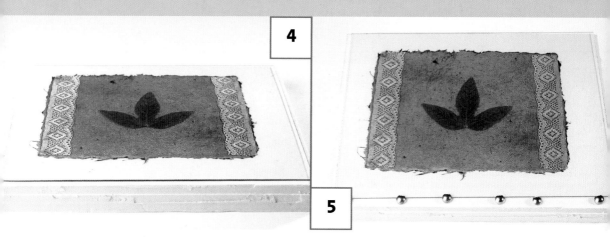

sunprint onto the sheet. Use straight pins on each corner of the paper to secure it, then pin the item in place, pushing the pins in deeply so they will not produce shadows. The time between applying the solution through the pinning should be no more than 15 minutes. You can store dry-coated paper in black, lightproof bags, but you must sunprint on it within a week.

4 If you wish to sunprint from negative photograph images on acetate transparencies, place a piece of glass over the transparency instead of pinning it in position. Support the glass with a row of T-pins at the bottom. The glass covering also works well for feathers and flat foliage.

5 Take the foamboard outside and prop it up so that it is perpendicular to the sun. Make sure it is secure, then leave for 10 minutes on a warm, sunny day and a little longer on a cool, sunny day. Do not leave for more than 20 minutes.

6 Bring the foamboard inside and remove any pins, glass, design objects, and transparencies. Rinse the sheet in a vat holding 3–4in (7.5–10cm) of cold water by gently moving it back and forth in the water for 10 seconds, design-side down. Holding the sheet carefully with one hand, empty and refill the container. Gently agitate the sheet again. Repeat until the water runs clear, usually 3–4 times.

7 Put the sheet on a drying board that has been sprayed with water, then sponge excess water from the sheet, removing any air bubbles. When the drying board is full of sheets, place it inside the house away from direct sunlight or drafts. If the sheets begin to buckle, cover them with paper towels and a heavy book, replacing the towels as they dampen, until the sheets are almost dry. Remove the sheets from the drying board when they are completely dry. Discard any unused solution immediately; do not store it for future use.

yellow-toned prints

After the blue-toned sheet has dried, turn the blue to yellow by rinsing the sheet in water and TSP. Mix 1tbsp (15ml) TSP into 1 quart (1 liter) of hot tap water; stir until the TSP is dissolved. Pour the solution into a vat, then put the sheet in the vat, design-side down, and gently move it back and forth until the blue color disappears. Immediately rinse the sheet in a vat of clear water, and dry it in the same way as a blue-toned sheet.

brown-toned prints

After the yellow-toned sheet has dried, turn the yellow to brown by rinsing the sheet in black tea. Steep 8–10 teabags in boiled water for 10 minutes. Remove the bags and pour the tea into a vat. Let it cool to a comfortable temperature, then put the sheet in the vat and gently move it back and forth until the yellow turns to brown. Immediately rinse the sheet in a vat of clear water, then dry as before.

Use the sunprinting technique to create blue, yellow, and brown designs on your papers.

Acknowledgment: For more information on this fascinating craft, see *Blueprints on Fabric* by Barbara Hewitt (see Resources, page 126).

cookie-cutter designs

Use cookie cutters to create paper shapes in a variety of designs, or to create a cutout shape in a sheet of paper. Use plastic cutters that will not rust, or spray metal cutters with cooking oil to help them resist the moisture.

paper shapes

Place the cookie cutters on the screen of the mold. Hold the mold over the vat of pulp, and with a meat baster, squirt the pulp evenly into the cookie cutters. Place the mold on a stack of newspapers and leave to drain for a few minutes to two hours. A longer draining time will allow the designs to keep their shape better. Carefully remove the cookie cutters, and lightly press the remaining shapes left on the mold with a towel to blot them. Flip the mold over onto a water-sprayed drying board, and lightly sponge the screen. Remove the mold and finish sponging the shapes until any excess water is gone. Use a toothpick to redefine edges if necessary. Dry in the usual way.

paper cutouts

Place a cookie cutter on a sheet of wet paper after it has been sponged on the drying board. Press the cookie cutter firmly and use a chopstick to remove the pulp from inside it. This will produce a deckled edge around the cookie-cutter shape, unlike cutting the design with scissors. To make a greeting card, fold the sheet with the cutout over a plain sheet of paper in a contrasting color.

paper shapes

paper cutouts

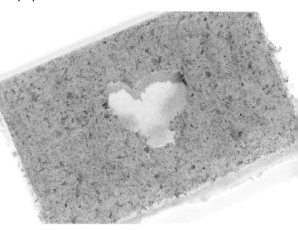

stitch designs

Sewing machine stitching is an innovative way to decorate handmade papers. Create your own designs, or trace them from embroidery books, catalogs, coloring books, or other sources. Always practice your design on bought paper first.

sewing a design

Lightly draw your design in pencil on the paper, then stitch over the pencil marks, using straight or fancy stitches. The tension should be slightly looser than for fabric so that the needle holes do not tear the paper; close stitches will also tear the paper. Erase any pencil lines and clip the thread ends close to the paper when you have finished. Apply a bit of glue to the ends on the back of the paper if your design will be subjected to stress.

You can also machine-stitch without any thread; the perforations will resemble those on punched tin pieces.

extra embellishments

To dangle buttons, confetti, or charms from the stitched design, stop the machine in the required place and pull a length of thread from the machine. The length you pull depends on how far you want the embellishments to dangle. Continue stitching and stopping wherever you want to dangle something. When you have finished the design, cut the pulled threads in half and tie on the embellishments.

sewing a design

extra embellishments

Acknowledgment: The designs illustrated here are by Andrea Niehuis of Amazing Yarns (see Resources, page 126).

resources & credits

north america

papermaking suppliers

Carriage House Paper
79 Guernsey Street
Brooklyn, NY 11222
Tel: (718) 599-PULP
          (800) 669-8781
Fax: (718) 599-7857

Coastal Fibers
1211 Marsh Street
Seal Rock, OR 97376
Tel: (541) 563-5459

Dieu Donné Papermill, Inc.
433 Broome Street
New York, NY 10013
Tel: (212) 226-0573
Fax: (212) 226-6088

Gold's Artworks
Lumberton, NC 28358
Tel: (800) 356-2306
Tel/Fax: (910) 739-9605

Green Heron Book Arts
1928 21st Avenue
Suite A
Forest Grove, OR 97116
Tel: (503) 357-7263
E-mail: bookkits@aol.com
Website: www.green-heronkits.com

Lee Scott McDonald
PO Box 264
Charlestown, MA 02129
Tel: (617) 242-2505
          (888) 627-2737
Fax: (617) 242-8825

Magnolia Editions
2527 Magnolia Street
Oakland, CA 94607
Tel: (510) 839-5268
Fax: (510) 893-8334

The Papertrail
170 University Avenue West
Waterloo
Ontario N2L 3E9
Canada
Tel: (800) 421-6826
          (519) 884-7123
Fax: (519) 884-9655

Twinrocker
PO Box 413
Brookston, IN 47923
Tel: (765) 563-3119
          (765) 563-3210
          (800) 757-8946
Fax: (765) 563-8946

Watson Paper Company
1719 Fifth Street NW
Albuquerque, NM 87102
Tel: (505) 242-9351
Fax: (505) 243-5644

dye companies

Dharma Trading Co.
PO Box 150916
San Rafael, CA 94915
Tel: (415) 456-7657
          (800) 542-5227
Fax: (415) 456-8747
E-mail: catalog@dharmatrading.com
Website: www.dharmatrading.com

Earth Guild
33 Haywood Street
Asheville, NC 28801
Tel: (828) 255-7818
          (800) 327-8448
Fax: (828) 255-8593
E-mail: inform@earthguild.com
          catalog@earthguild.com

earthues
A Natural Dye Company, Inc.
5129 Ballard Avenue NW
Seattle, WA 98107

Tel: (206) 789-1065
Fax: (206) 783-9676
Sells a natural dye instruction booklet

Guerra Paint & Pigment
510 East 13th Street
New York, NY 10009
Tel: (212) 529-0628
Fax: (212) 529-0787

PRO Chemical & Dye, Inc.
PO Box 14
Somerset, MA 02726
Tel: (888) 2-BUY-DYE
Fax: (508) 676-3980

Richters
357 Hwy 47
Goodwood
Ontario L0C 1A0, Canada
Tel: (905) 640-6677
          (800) 668-4372
E-mail: orderdesk@richters.com
Website: www.richters.com
Large selection of natural dye plants and seeds

other helpful resources

Amazing Yarns
2559 Woodland Place
Redwood City, CA 94062
Tel: (650) 306-9218
E-mail: amazingyarns@yahoo.com
Large selection of unusual yarns

Blueprints on Fabric by Barbara Hewitt
ISBN 0-934026-91-2
Interweave Press, Inc.
201 East Fourth Street
Loveland, Colorado 80537

Blueprints-Printables
1129 Cortez Avenue
Burlingame, CA 94010
Tel: (800) 356-0445
E-mail: cyanoprint@aol.com
For sunprinting chemicals

Loose Ends
PO Box 2031
Salem, Oregon 97307
Tel: (503) 390-7457
Fax: (503) 390-4724
Website: www.Loosends.com
Natural fiber trims

Michaels
The Arts and Crafts Store
8000 Bent Branch Drive
Irving, Texas 75063
Tel: (972) 409-1300
Website: www.michaels.com
For general craft supplies

united kingdom

Fred Aldous Ltd
37 Lever Street
Manchester M60 1UX
Tel: (0161) 236-2477
General craft supplies

B&Q
Portswood House
1 Hampshire Corporate Park
Chandlers Ford
Hampshire SO53 3YX
Tel: (02380) 256-256
Website: www.diy.com
General hardware, including timber,
brass mesh, laminated boards, acetate sheets,
and adhesives

Falkiner Fine Papers Ltd
76 Southampton Row
London WC1B 4AR
Tel: (020) 7831-1151
Fax: (020) 7430-1248
Partially processed fiber, size, and general
art and craft supplies

Homebase
Beddington House
Railway Approach
Wallington
Surrey SM6 0HB
Tel: (020) 8784-7200
Website: www.homebase.co.uk
General hardware, including timber,
brass mesh, laminated boards, acetate sheets,
and adhesives

Home Crafts Direct
PO Box 38
Leicester LE1 9BU
Tel: (0116) 251-3139
Fax: (0116) 251-5015
Website: www.homecrafts.co.uk
Complete range of papermaking supplies
available by mail order, including
partially processed fiber, size, molds
and deckles, vats, dyes and other colorants,
and sunprinting materials

T N Lawrence & Son Ltd
117–119 Clerkenwell Road
London EC1R 4AR
Tel: (020) 7242-3534

Fax: (020) 7430-2234
Email: artbox@lawrence.co.uk
Website: www.lawrence.co.uk
Partially processed fiber and general
art supplies

John Lewis
Oxford Street
London W1A 1EX
Tel: (020) 7629-7711
Website:
         www.johnlewispartnership.co.uk
Dyes, yarns, and general art and
craft supplies

Paperchase
213 Tottenham Court Road
London W1P 9AF
Tel: (020) 7467-6200
Foamboard, stencil plastic, adhesives,
general art and craft supplies, and a
wide range of special papers and
other stationery

Specialist Crafts
PO Box 247
Leicester LE1 9QS
Tel: (0116) 251-0405
Fax: (0116) 251-5015
Email: post@speccrafts.co.uk
Website: www.speccrafts.co.uk
Complete range of papermaking supplies
available by mail order, including partially
processed fiber, size, molds and deckles, vats,
dyes, and other colorants

author's acknowledgments

I am deeply grateful to the following people: Nancy Nehring for her continued inspiration and assistance and for her example in all book matters; Paul Martin for his support of this project and for so generously sharing his expertise on natural dyeing; Andrea Niehuis for her artful sewing-machine techniques and help in developing the stitching embellishment; Barbara Hewitt for her crash course in blueprinting and for proofing; Heather Lockie and Shawn Lockie for being my papermaking muses; Bob Goligoski for his unconditional support and for graciously living in an "artist colony" for the four years in which I wrote the book; Kate Kirby at Collins & Brown for anchoring the book from start to finish; Michelle Pickering and Jane Ellis, who heroically manned it in midstream and pounded out the dents; Liz Brown, Alison Lee, and Sian Irvine for polishing it, beautiful beyond expectation.

index